Diego Ricardo Gutiérrez
Manuel Alberto Sandez
Lucas Mansilla

MANUAL OF ARTISANAL GOAT DAIRY PRODUCTS

Diego Ricardo Gutiérrez
Manuel Alberto Sandez
Lucas Mansilla

MANUAL OF ARTISANAL GOAT DAIRY PRODUCTS

Guide to improve the production systems of different products made from goat milk

ScienciaScripts

Imprint

Any brand names and product names mentioned in this book are subject to trademark, brand or patent protection and are trademarks or registered trademarks of their respective holders. The use of brand names, product names, common names, trade names, product descriptions etc. even without a particular marking in this work is in no way to be construed to mean that such names may be regarded as unrestricted in respect of trademark and brand protection legislation and could thus be used by anyone.

Cover image: www.ingimage.com

This book is a translation from the original published under ISBN 978-3-639-53536-5.

Publisher:
Sciencia Scripts
is a trademark of
Dodo Books Indian Ocean Ltd. and OmniScriptum S.R.L Publishing group
Str. Armeneasca 28/1, office 1, Chisinau-2012, Republic of Moldova, Europe
Printed at: see last page
ISBN: 978-620-5-31652-8

Copyright © Diego Ricardo Gutiérrez, Manuel Alberto Sandez, Lucas Mansilla
Copyright © 2022 Dodo Books Indian Ocean Ltd. and OmniScriptum S.R.L Publishing group

MANUAL OF ARTISANAL GOAT DAIRY PRODUCTS

Artisanal goat dairy products manual

Authors:

Diego Ricardo Gutierrez
Food Engineer, Ph. D.

Manuel Alberto Sandez
Agricultural and Food Industries Eng.

Lucas Mansilla
Agricultural and Food Industries Eng.

Editors

Dr. Eng. Diego Ricardo Gutiérrez

Manuel Alberto Sandez

Lucas Mansilla

Eng. Adrián Suarez

Eng. Edmundo Vizgarra Gómez

CONTENTS

Introduction

Goat milk in terms of its composition is very similar to breast milk, it is easier to digest than cow's milk, has a high content of essential amino acids and fatty acids, can be used by those who are lactose and casein intolerant because of its lower content of these substances. Due to its high calcium value, its consumption is recommended for older adults and young children, mainly to prevent osteoporosis. For those who have difficulties in ingesting cow's milk, goat's milk represents an alternative to consider.

However, since it is the most nutrient-rich natural food, it is quickly degraded by microorganisms, which makes it necessary to take all precautions to prevent this from happening, either through pasteurization or through the elaboration of different dairy by-products. This industrialization would benefit small and medium-sized producers, most of whom only use it as a means of family subsistence, not only nutritionally but also economically.

The purpose of this publication is to give indications on good manufacturing practices to be taken into account in all manufacturing processes, remembering that a good quality product is obtained from raw material with the same conditions.

Milk

According to the Argentine Food Code, milk is defined as the product obtained from the total and uninterrupted milking of a dairy animal that is in optimal health and feeding conditions.

The name milk without further clarification applies only to milk that comes from the milking of the cow, when it comes to milk of another species should be clarified which species it is, for example: goat's milk.

Composition

Tabla 1. Composición de la leche de cabra

Composición de la leche de cabra (%)	
Sólidos totales	11,70-15,21
Proteína (Nx6,38)	2,90-4,60
Grasa	3,00-6,63
Lactosa	3,80-5,12
Cenizas	0,69-0,89
pH	6,41-6,70

Fuente: Boza *et al.*, 1992, citado por Cruz *et al.*, 2012

Properties of goat milk

- Greater digestibility: the organism absorbs and digests it more easily than other milks of other species.
- Contains proteins of high biological value: in addition to its high biological value, it provides about 30 g of proteins.
- Prevents osteoporosis: it is an important source of calcium and vitamin D, so it is recommended for children and older adults.
- Improves the recovery of people with anemia: it has the property of regenerating α-hemoglobin.

Comparison of milk composition of different species:

Composición	Cabra	Oveja	Vaca	Humana
Grasa %	3.8	7.9	3.6	4
Sólidos no Grasos %	8.9	12	9	8.9
Lactosa %	4.1	4.9	4.7	6.9
Proteína %	3.4	6.2	3.2	1.2
Caseína %	2.4	4.2	2.6	0.4
Albumina, globulina %	0.6	1	0.6	0.7
N no proteico %	0.4	0.8	0.2	0.5
Cenizas %	0.8	0.9	0.7	0.3
Calorías/100 ml	70	105	69	68

Fuente: Park (2006)

4. Milk quality

4.1. Quality from the Hygienic Point of View

To be acceptable, a milk must have:

➢ Good conservation capacity
➢ It must be free of pathogenic germs.
➢ High nutritional value
➢ Be clean, free of foreign matter and sediments.
➢ Do not contain chemical residues (pesticides).

4.2. Quality from the technical point of view

Good quality raw milk:

➢ Must not be tasteless or have abnormal color or odor.
➢ Must have a low bacterial content

- ➢ Must have a normal composition
- ➢ It must present values of acidity, density and
- ➢ Normal fat content.

4.3. Milk quality as a function of udder condition

Healthy alveolus:

- ➢ Healthy udder cistern
- ➢ Contamination-free milk of good quality
- ➢ Wash with water plus disinfectant
- ➢ Healthy and clean nipple free of fecal matter, soil or dirt and undamaged.

Higher production and good quality milk are obtained.

Sick alveolus:

- ➢ Sick udder cistern (mastitis)
- ➢ Milk with presence of: somatic cells, streptococci, coliforms. Not accepted for consumption
- ➢ Microorganisms from the soil, guano and environment that contaminate the milk through the teat.
- ➢ Sick and contaminated canal

Lower production and poor quality milk

5. BASIC ANALYSIS FOR MILK QUALITY CONTROL

5.1. Sample Collection

The sample must be representative so that the results are as close to reality as possible; to achieve this, the size and shape of the container containing the product, the uniformity and, finally, the type and time of agitation or mixing must be considered. When the volume of the containers is large, they should be agitated with suitable mechanical means for 5 to 10 minutes. When the sample is to represent the contents of several jars, take two samples for each 2 to 5 jars, three for 6 to 60 jars, four for 61 to 80, five for 81 to 100 and finally one more sample for each additional 100 jars or fraction thereof.

5.2.3. Determination of Organoleptic Properties

a- Determination of color: Fresh milk is yellowish white, with a certain cream color when it is very rich in fat. Skim milk or milk very low in fat content is white with a slightly bluish tint.

b- Determination of flavor: Fresh milk has a slightly sweet flavor, caused by its high lactose content. Due to its feeding, it easily acquires the flavor of the pastures it may consume, a very well known example is the consumption of clover.

c- Determination of odor: When the milk is fresh it has almost no characteristic odor, but it easily acquires the aroma of the containers in which it is kept; a small acidification already gives it a special odor, as does contact with strong-smelling elements.

5.2. Basic Analysis

5.2.1. Acidity Analysis Acidity analysis can be used to determine not only the commercial quality of the milk, but also the microbiological quality and the handling of the milk during the extraction, transport and conservation process. According to the Argentine Food Code, the acidity value that can be considered normal in goat milk should be between 12 and 16 expressed as degrees Dornic. A higher value would indicate a degradation of lactose caused by

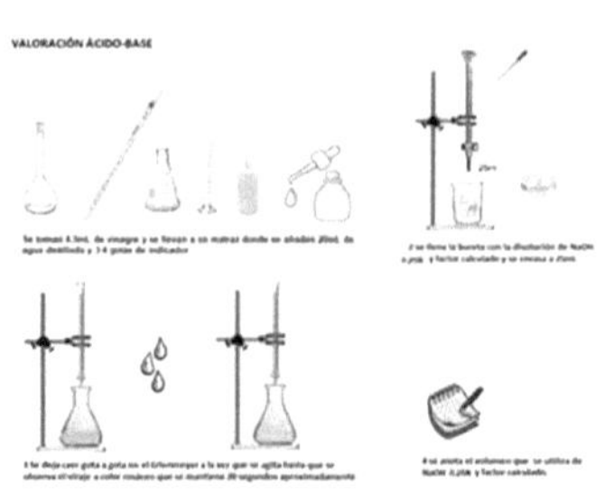

Figure 1. Schematic diagram of titration realization

the development of lactic bacteria, this degradation affects the stability of proteins, which impairs the coagulation capacity of milk and therefore the final yield of cheeses or rapid coagulation when heating occurs.

The method used to determine the acidity value is very simple, using a 25 ml burette, a 250 ml Erlenmeyer flask, Dornic solution (sodium hydroxide solution of 0.1N concentration) and phenolphthalein as an indicator.

The procedure is as follows: the burette is filled with the sodium hydroxide solution, 10 ml of milk is placed in the Erlenmeyer and 4 drops of the indicator solution are added. The hydroxide solution is dropped slowly and drop by drop, keeping the milk moving to ensure good mixing. The end point is determined when a pale pink coloration appears and remains for more than 30 seconds, at which time the dripping of

Figure 1. Milk titration

the hydroxide solution is stopped and the volume that has been used is noted. The results are obtained by converting the spent milliliters of hydroxide solution to Dornic degrees as follows:

Figure 3. Milk titration equipment

1 ml of Sodium Hydroxide solution = 10 degrees Dornic acidity.

Alternatives to measure acidity when laboratory elements are not available to determine the value of acidity in milk, an approximate idea can be obtained by using strips of pH paper. It is necessary to remember that the pH value for good quality milk is 6.6 and that values from 6 to 6.8 are equivalent to acidity values in Dornic degrees between 14 and 20 degrees, which are considered acceptable for use in cheese production. The method is simple, it consists of introducing the strip of paper

Figure 2. Density measurement in goat milk

in a glass with the milk sample for about 10 seconds, when extracting it and by comparison of the color of the strip the approximate value of pH is determined.

Table 1. pH values expressed in degrees Dornic.

pH value	Value in Dornic degrees
6,5	

5.2.2. Determination of the Density value It allows to know if water was added to the milk with the intention of increasing its volume. The determination is simple and

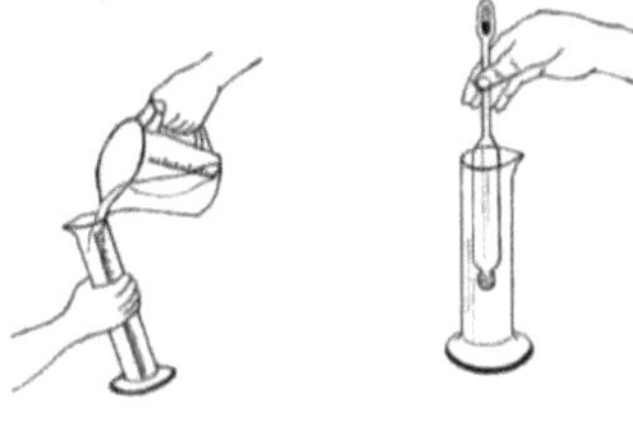

Figure 3. Scheme for density measurement with lactodensimeter.

easy to interpret. The elements used are a 250 mL test tube, a lactodensimeter and a temperature correction table. The necessary procedure to carry out this analysis is the following: the burette is filled up to the mark of 250 mL and in the center of the same one the lactodensimeter is introduced slowly and with rotating movements, when remaining still or without movement it is observed where the edge of the milk and the stem of the lactodensimeter are put in contact and the number that is observed and the temperature at which this reading is made is noted down. With the measurement taken and the temperature value recorded, it is interpolated in the correction table for 15 degrees Celsius and the definitive value of the density for that milk sample is obtained. If the resulting value is not between 1.028 and 1.033 grams per milliliter at a temperature of 15 degrees Celsius, it is concluded that the milk may have suffered the addition of water, which is not allowed and all the milk may be rejected.

Figure 1. Images of goat milking pallets.

Obtaining goat milk

6.1. Milking facilities

Some examples of constructions using wooden pallets are shown below.

6.2. Milking routine

6.2.1. Before milking

Monitor udder health on a regular basis

➢ Always use disposable gloves and make sure they are clean.
➢ Check udder health on a regular basis; first check newly calved first-calving goats and those with 1 - 2 weeks post parturition.
➢ Keep a record of udder health results for each animal.
➢ Always keep the milk of sick goats separate.

Follow the correct milking order

➢ Start by milking the healthy goats first and the healthy first calving goats first.
➢ Always milk the oldest and sickest goats last or separately.

Regularly draw the first 3 squirts of milk into a dark bottomed jug.

➢ Always wear disposable gloves and make sure they are clean.
➢ Place 2-3 squirts of milk from each nipple in a blunt jug.
➢ Examine the milk for clots, color changes or other inconsistencies.
➢ Always store abnormal milk separately.
➢ The blunting also stimulates milk letdown

Disinfect and dry teats before milking.

➢ Always wear disposable gloves and make sure they are clean.
➢ Clean each nipple using paper towels or napkins.

> It is recommended to disinfect teats before milking: use a disinfectant approved for teat and milk contact; wait 30 seconds before removing it.
> In all cases use disposable paper to carefully clean and dry each udder.
> Never use the same paper or towel for more than one goat.

6.2.2. After milking

Seal to disinfect teats immediately.

> Always use disposable gloves and make sure they are clean.
> Completely dip teats as soon as milking is completed. Use an approved disinfectant sealant, as this is the most effective way to prevent the spread of mastitis.
> Keep goats standing for at least 15 minutes after milking.
> Ensure proper cooling of milk
> Always check temperatures to ensure that the correct cooling is achieved during and after milking.
> Always be sure to follow the specific cooling temperature recommendations for milk.

Regular monitoring of milking results

> Ask for and review the information sent by the establishment where the milk is industrialized, mainly regarding the quality and composition of the milk on a regular basis.

Cleaning of utensils

> Milk storage and transport garbage cans need to be cleaned immediately after emptying, using hot water and approved detergents.
> Rinse well with warm water and dry until the next day.
> Before starting the next day rinse all the elements with cold water and let drain until use.

Figure 4. Teat sealing scheme

Teat sealing Teat disinfection is an essential step in the prevention of udder infections. It serves to reduce the number of bacteria present on the teat skin. It is carried out at two specific times during the milking process: before and after milking. Cleaning and drying the teats before milking is, on the other hand, essential to complete the disinfection, in addition, this stage stimulates the milk let-down reflex. It is recommended to prepare the sealant solution by mixing one liter of iodine solution with one liter of household detergent. The application method is to introduce the entire clean nipple into the sealant which is placed inside a glass or container containing it for approximately 30 seconds.

6.4. Mastitis control milk quality has a direct relationship with high somatic cell counts (SCC), which are related to mastitis, and is closely related to milking practices. It is identified as the inflammation of one or more teats of the udder. The bacteria that cause mastitis enter through the teat canal, penetrate the milk-producing cells and multiply in them. In order to detect mastitis, the California Mastitis Test (CMT) must be performed.

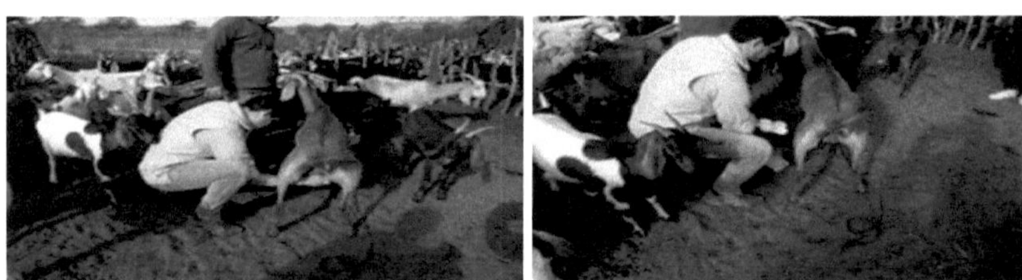

Figure 5. Milking and sample collection for mastitis test.

Steps to follow for mastitis detection:

Discard the first 2 streams from each teat and milk 2 mL of milk onto the test paddle.

Figure 6. Mastitis detection

Place 2 ml of reagent and move the paddle in circles to mix. Keep moving and observe after 15 seconds, note the results.

When is the California Test used?

Monthly, in all animals in production. And when symptoms of mastitis are suspected, such as swollen udders and higher than normal temperatures.

What are the results?

- ✓ Level 1: no change or no reaction.
- ✓ Level 2: presence of some mucus-like filaments.
- ✓ Level 3: presence of abundant number of strands
- ✓ Level 4: Mucus sticking to the pallet

How to interpret the results:

Table 2. Incidence level of mastitis in milk

Incidence level of mastitis	What is done with the milk?
Low Risk: 1 and 2	Suitable for consumption and processing of dairy products.
High Risk: 3 and 4	Not suitable for consumption.

4. Milk hygiene

4.1. Milk cooling

The multiplication of germs in milk is prevented by cooling the freshly milked product to temperatures below 10 °C and keeping it under these conditions until it is used in the industry. Milk cooling can be done economically with natural fresh water or with artificially cooled water (by adding ice or also urea). At the time of extraction, the milk is at a temperature of 37 °C. In winter this value can be reduced by immersing the drums containing the milk in running water, reducing it by up to 10°C.

In summer or hot weather, milk drums can be immersed in cold water, the water can be cooled in two ways:

➢ Adding ice and a little salt to the container containing water.
➢ Dissolving Urea in the water to be used for cooling, using 1 kg of urea per 10 liters of water.

This method reduces the temperature of the milk by more than 15 °C, thus ensuring hygienic quality from the microbiological point of view.

4.2. Conservation

Milk should be stored at temperatures below 10 °C if not used immediately after milking. It is recommended to use stainless steel or plastic containers with lids. **Milk should not be frozen**, because it affects the coagulation properties of the proteins and consequently reduces the yield of the processing process.

Method of goat cheese production

Artisanal production of goat cheeses

Definition According to the FAC, goat cheese is the product obtained by coagulating, salting and maturing goat's milk, to which lactic ferment, calcium chloride and permitted substances may be added to give a specific characteristic to the final product.

5.2. Dairy quality for cheese production

- Healthy animals ------> Herd health checks
- Health calendar
- Healthy and balanced nutrition -----> To satisfy maintenance and production requirements Without altering the composition, taste and odor of the milk.
- Hygienic milking ------ > Correct hand or mechanical milking
- Hygiene of the gland. Mammary gland, nipple, utensils, and personnel.
- Cleaning of the facilities
- Milk purity ------ > Avoid contamination by dirty utensils, foreign bodies, insects, pesticides, disinfectants, detergents, antibiotics.

5.3. Ingredients

Ingredients to make goat cheese:

- Freshly milked milk
- Calcium hydroxido (optional)
- Rennet
- Lactic ferment (optional).
- Calcium chloride.
- Salt die (coarse salt + water).

5.4. Utensils

Utensils necessary for the elaboration:

- A large stainless steel pot
- Canvas, perfectly clean, approximately 1.5 x 1.5 meters (5 x 5 feet)
- 1 Thermometer that reaches a minimum of 65 °C (104 °F)
- 1 Lira or a sharp knife.
- Cheese molds.
- Brine storage container
- Shovel or wooden spoon.

5.5. Hygienic-sanitary practices

- Hygiene of premises and personnel
- To begin with, it is of vital importance to thoroughly clean the workplace, the elements to be used and the hands.
- Particular attention should be paid to the circumstances in which milk processing takes place in rural areas, where there is no space adequately prepared for this.
- A place intended for the treatment of milk should, whenever possible, be protected against the entry of flies and insects. Work tables and utensils should not be made of wood, but if they are, they need to be cleaned and disinfected correctly and permanently.
- Places should be well ventilated and have the necessary hot water to disinfect all utensils and containers in contact with the milk every day.

Only impeccable cleaning avoids the danger of foreign contamination.

Rules for proper cleaning:

- Thoroughly rinse utensils immediately after use. To remove coarse soiling, then wash with a detergent.
- Use suitable brushes and sponges for cleaning.
- After cleaning, rinse with plenty of lukewarm water, removing all traces of cleaning products.
- Check with the eye the good state of cleanliness and remove possible dirt.
- Once the containers have been cleaned, they should be turned upside down and left to dry, thus depriving the microorganisms of the water they need to grow.

Goat cheese production process

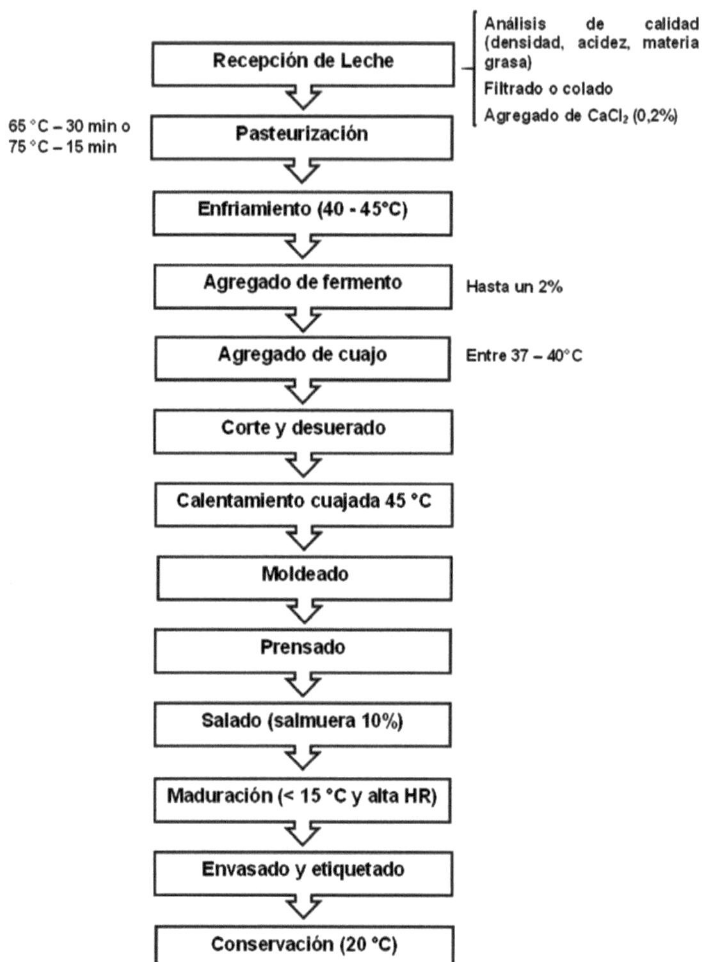

Figure 7. Block diagram of the goat cheese production process.

Description of the processing stages

5.5.1. Milk reception Milk analysis is carried out as described above.

Figure 8. Milk analysis

Strain the milk over the container (pot). This step is important because, whether milked by hand or with a milking machine, it is normal for the milk to have some goat hair or other type of dirt.

Figure 9. Milk straining

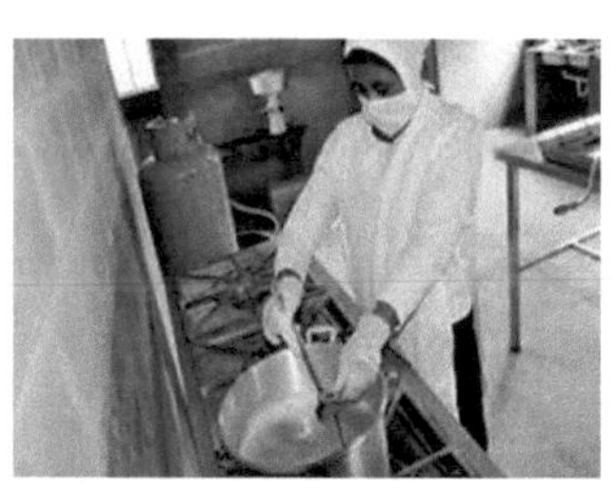

Figure 10. Pasteurization of milk

Pasteurization the next step is to pasteurize by keeping the milk at 65 °C for 30 minutes, this stage is important to ensure that we eliminate risks, by keeping the milk at a certain temperature for a certain time, we ensure the death of certain bacteria and pathogenic microorganisms, such as those that cause the dreaded brucellosis.

Once the milk has been pasteurized, the temperature should be lowered, just wait until it reaches a value of 40 °C.

If a lactic **starter** is used, it is added at this time in an amount equal to or less than 2%; whey extracted during the previous day's processing and kept in a cool environment can be used.

Figure 11. Milk cooling

Lactic acid bacteria, which in these cases we can acquire from yogurt, provide us with good benefits because:

- They produce acid, which is necessary for the formation of the coagulum when rennet is added and helps the curd to drain, leaving a compact mass.

- It prevents and inhibits the development of pathogenic and spoilage-causing microorganisms.
- They are responsible for the flavor and texture of cheeses.

Figure 12. Aggregate of ferment

When the temperature reaches 37 °C it is time to add **rennet.** There are several ways of renneting milk. We are going to make an enzymatic fermentation, by adding two enzymes pepsin and chymosin, which act on the casein or milk protein precipitating it, giving rise to curd. For this we are going to use rennet. Rennet is a rennin enzyme extracted from the fourth stomach of lactating ruminants, however, given the demand, there are commercially processed rennets on the market, both of animal and vegetable origin, as well as

Figure 13. Rennet aggregate

microbial, usually liquid rennet is used.

The quantity to be used will depend on the rennet titer, we understand as titer the concentration, the most common is to use rennet with a titer of 1/1000, that is to say, one liter of rennet has the capacity to coagulate 1000 liters of milk. It is also possible to use rennet powder, which is sold in pharmacies, and calculate the proportion to be added according to the volume of milk. Before using rennet powder, dissolve it in a small glass of water with 1 teaspoon of fine salt. It is also possible to use non-commercial rennet prepared at home or by hand, as described in the recommendations.

Then add the rennet to the milk and stir, leave to stand, maintaining the temperature, for approximately 35-40 minutes, at which point we will see how the surface inside the pot becomes covered with an increasingly compact mass, i.e. the curd is formed.

Figure 14. Curd heating

Cutting the curd is done using the lyre, but at home we can use a knife with a long sharp blade, cut the curd into squares and move the pot in a slight back and forth motion for 10 minutes, this helps the curd to drain.

Figure 15. Curd cutting

Heating the curd when a more compact curd and consequently a firmer cheese can be obtained by heating the curd together with the whey at a temperature of 50 °C and keeping it in a slow rotating movement for no more than 15 minutes.

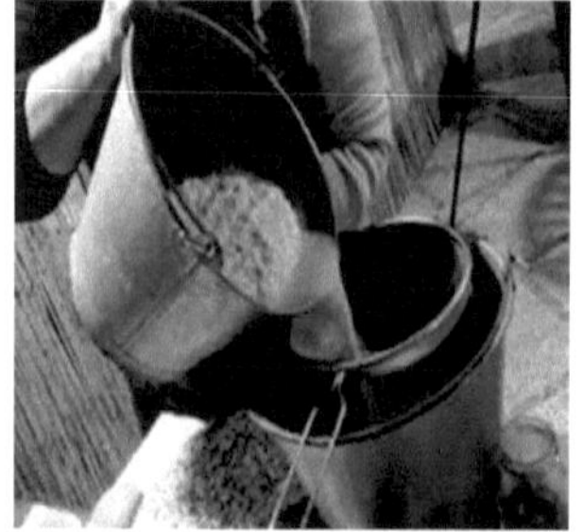

Separation of the Curd now we separate the curd from the whey, using for this purpose the canvas. Let the curd drain for a few minutes, after which, taking the ends of the cloth, we make a kind of bag, squeeze it and force the whey out even more.

Figure 16. Curd separation

Molding the next step consists of cutting the curd and placing it in the molds, taking care to cover the molds with a cloth if plastic

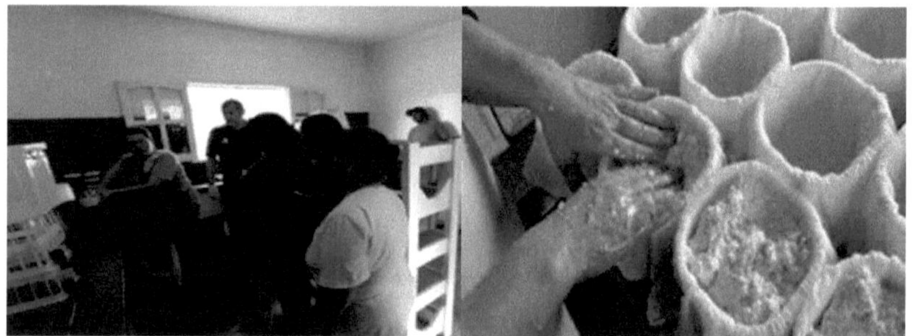

Figure 17. Curd molding

containers are used; if stainless steel containers are used, this is not necessary.

Pressed we put a weight on top and let it drain for a couple of hours. After that time, we remove the cheese from the mold, separate the canvas and wash it to be used again, it is generally recommended to press with a weight equal or equivalent to four times the final weight that is expected to be obtained, it is also important to rotate the position of the cheese in the mold to keep the shape, well define the edges of the mold.

Figure 18. Curd pressing

Figure 19. Cheese salting

The cheese is placed in brine (salt + water) which is prepared by dissolving one kg of common salt in one liter of water, this solution should be at a temperature not exceeding 15 ºC. It should be kept in the brine for at least 3 hours, taking care that the cheeses are always submerged, in case they float the surface should be covered with solid common salt, it is extremely important to keep the cheese cold at all times.

Once the recommended time for salting has elapsed, the cheeses go on to the last stage of processing, where all the organoleptic characteristics (flavor, aroma, color, texture) are defined. For this to happen, they should be placed in a cold environment with temperatures below 15 °C, with low humidity and scarce air circulation. The total ripening time will depend on the

Figure 21. Cheese ripening

type or variety of cheese to be obtained.

Once ripening is complete, the cheese is ready for marketing and can be vacuum packed in wooden boxes, cardboard boxes, wrapped with paper, etc.

Figure 20. Packaged cheese and labeling

Conservation keep the cheeses in conditions until the moment of sale at low temperature and high Relative Humidity (to avoid weight loss).

Obtaining a healthy, safe and high quality product requires the control and monitoring of each stage of the production process, from

Figure 22. Cheese preservation

feeding the herd, obtaining the milk, handling, processing and maturing of the cheeses.

Recommendations

- **For cleaning and disinfection** Utensils and all parts that come into contact with milk can be disinfected with hot water at 80 ºC before use. A simple and water-saving method is to fill a tub with hot water and put directly into it all the utensils necessary for the treatment of milk. They are then removed directly from the hot water for use. During the cheese making process, the water cools down a little, but remains practically sterile. If visibly dirty and undisinfected utensils are put into the slightly cooled water, they should be replaced with fresh hot water.

- **For the addition of the ferment, allow the** necessary time for the lactic acid bacteria to be able to form lactic acid from lactose, the milk sugar. The appropriate rest is about 45 minutes.
- **After adding the rennet**, it is necessary to let the milk with the rennet stand for at least 45 minutes, during which time the container in which the coagulation is taking place should not be moved for any reason, nor should it be mixed.

- **Use of non-commercial rennet**

Extraction of rennet from kid

- Slaughtering a kid in a place sheltered from wind and earth.

- Remove and empty the rennet (or stomach), wash it with potable water and remove all the fat.

- Place the clean curd in coarse salt, covering it completely inside and out.

- Dry in the shade and protected from dust, for 3 days or more, until dry.

- Store in a cool place (refrigerator) and protected from light until use. It should not have unpleasant odor at the moment of use.

- **Use of rennet**

- Heat for 15 minutes 1 liter of fresh whey (from the last cheese making) to 70° C and then cool to 40° C (lukewarm).

- Place the serum in a wide-mouth glass bottle, well cleaned.

- Add the dried curd in small pieces and let stand for 2 or 3 days at room temperature, in a dark place. Then keep in the refrigerator. Use the preparation without adding whey or rennet again. It should not have an unpleasant odor at the time of use.

- **How to measure rennet titer**

- Place 1 teaspoon (1 cubic centimeter) of rennet in 100 cubic centimeters of milk (half a cup of tea).

- Leave to act for 15-20 minutes, during which time it should coagulate.

- If it does not coagulate in time, double the amount of rennet should be used to achieve the result, testing again.

- **For pressing**

It is advisable to use as pressing weight a value equal to that of the product being processed, for example, for each kilogram of cheese, use one kilogram of pressing.

- **Importance of salting**

It is suggested to carry out a correct salting because this is the way to achieve it:

- Improve flavor.

- Control the development of microorganisms.

- In some cases, if necessary, it helps to form the bark or shell.

Method of goat cheese production

Artisanal production of goat cheeses

Definition According to the FAC, goat cheese is the product obtained by coagulating, salting and maturing goat's milk, to which lactic ferment, calcium chloride and permitted substances may be added to give a specific characteristic to the final product.

5.2. Dairy quality for cheese production

- Healthy animals ------> Herd health checks
- Health calendar
- Healthy and balanced nutrition -----> To satisfy maintenance and production requirements Without altering the composition, taste and odor of the milk.
- Hygienic milking ------ > Correct hand or mechanical milking
- Hygiene of the gland. Mammary gland, nipple, utensils, and personnel.
- Cleaning of the facilities
- Milk purity ------ > Avoid contamination by dirty utensils, foreign bodies, insects, pesticides, disinfectants, detergents, antibiotics.

5.3. Ingredients

Ingredients to make goat cheese:

- Freshly milked milk
- Calcium hydroxide (optional)
- Rennet
- Lactic ferment (optional).
- Calcium chloride.
- Salt die (coarse salt + water).

5.4. Utensils

Utensils necessary for the elaboration:

- A large stainless steel pot
- Canvas, perfectly clean, approximately 1.5 x 1.5 meters (5 x 5 feet)
- 1 Thermometer that reaches a minimum of 65 °C (104 °F)
- 1 Lira or a sharp knife.
- Cheese molds.
- Brine storage container
- Shovel or wooden spoon.

5.5. Hygienic-sanitary practices

- Hygiene of premises and personnel
- To begin with, it is of vital importance to thoroughly clean the workplace, the elements to be used and the hands.
- Particular attention should be paid to the circumstances in which milk processing takes place in rural areas, where there is no space adequately prepared for this.
- A place intended for milk processing should, whenever possible, be protected against the entry of flies and insects. Working tables and utensils should not be made of wood, but if they are, they need to be cleaned and disinfected correctly and permanently.
- Places should be well ventilated and have the necessary hot water to disinfect all utensils and containers in contact with milk every day.

Only impeccable cleaning avoids the danger of foreign contamination.

Rules for proper cleaning:

- Thoroughly rinse utensils immediately after use. To remove coarse soiling, then wash with a detergent.
- Use suitable brushes and sponges for cleaning.
- After cleaning, rinse with plenty of lukewarm water, removing all traces of cleaning products.
- Check with the eye the good state of cleanliness and remove possible dirt.
- Once the containers have been cleaned, they should be turned upside down and left to dry, thus depriving the microorganisms of the water they need to grow.

Goat cheese production process

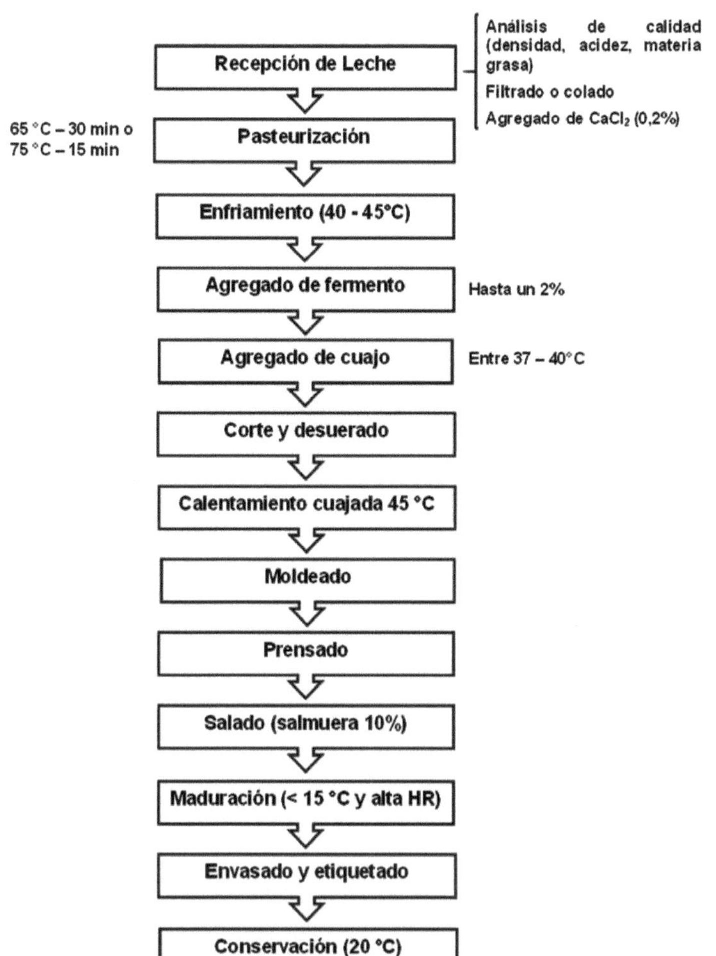

Figure 23. Block diagram of the goat cheese production process.

Description of the processing stages

5.5.1. Milk reception Milk analysis is carried out as described above.

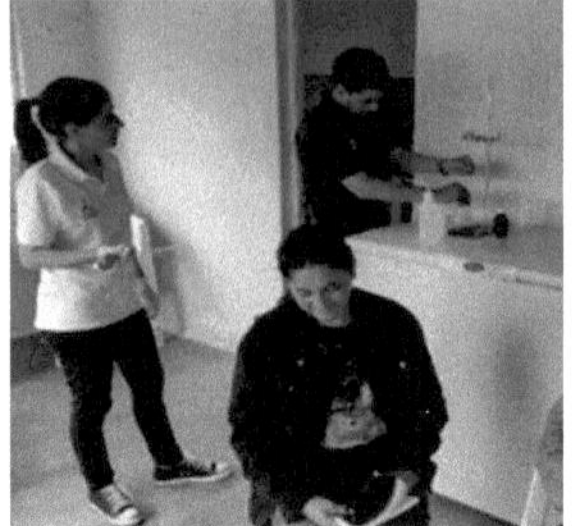

Figure 24. Milk analysis

Strain the milk over the container (pot). This step is important because, whether milked by hand or with a milking machine, it is normal for the milk to have some goat hair or other type of dirt.

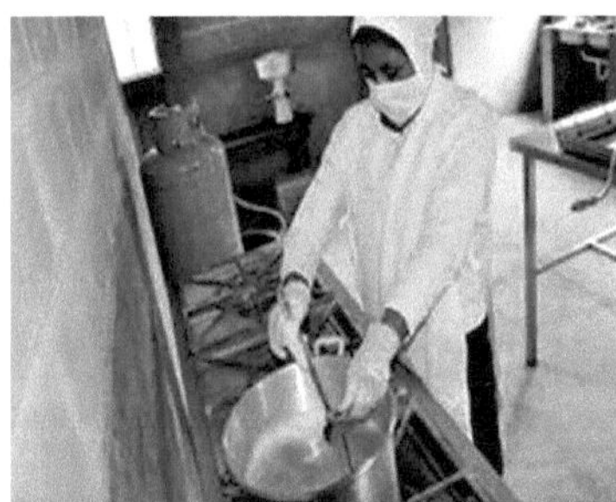

Figure 26. Pasteurization of milk

Pasteurization the next step is to pasteurize by keeping the milk at 65 °C for 30 minutes, this stage is important to ensure that we eliminate risks, by keeping the milk at a certain

Figure 25. Milk cooling

temperature for a certain time, we ensure the death of certain bacteria and pathogenic microorganisms, such as those that cause the dreaded brucellosis.

Figure 27. Milk straining

Once the milk has been pasteurized, the temperature should be lowered, just wait until it reaches a value of 40 ºC.

If a lactic **starter** is used, it is added at this time in an amount equal to or less than 2%; whey extracted during the previous day's processing and kept in a cool environment can be used.

Lactic acid bacteria, which in these cases we can acquire from yogurt, provide us with good benefits because:

Figure 28. Aggregate of ferment

- They produce acid, which is necessary for the formation of the coagulum when rennet is added and helps the curd to drain, leaving a compact mass.
- It prevents and inhibits the development of pathogenic and spoilage-causing microorganisms.
- They are responsible for the flavor and texture of cheeses.

When the temperature reaches 37 °C it is time to add **rennet.** There are several ways of renneting milk. We are going to make an enzymatic fermentation, by adding two enzymes pepsin and chymosin, which act on the casein or milk protein precipitating it, giving rise to curd. For this we are going to use rennet. Rennet is a rennin enzyme extracted from the fourth stomach of lactating ruminants, however, given the demand, there are commercially processed rennets on the market, both of animal and vegetable origin, as well as microbial, usually

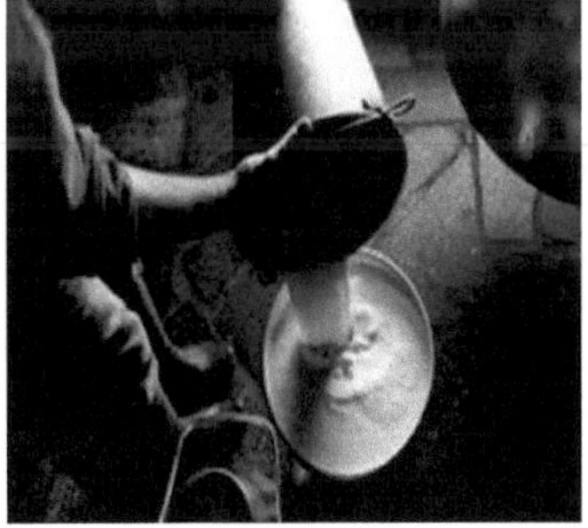

Figure 29. Rennet aggregate

liquid rennet is used.

The quantity to be used will depend on the rennet titer, we understand as titer the concentration, the most common is to use rennet with a titer of 1/1000, that is to say, one liter of rennet has the capacity to coagulate 1000 liters of milk. It is also possible to use rennet powder, which is sold in pharmacies, and calculate the proportion to be added according to the volume of milk. Before using rennet powder, dissolve it in a small glass of water with 1 teaspoon of fine salt. It is also possible to use non-commercial rennet prepared at home or by hand, as described in the recommendations.

Then add the rennet to the milk and stir, leave to stand, maintaining the temperature, for approximately 35-40 minutes, at which point we will see how the surface inside the pot becomes covered with an increasingly compact mass, i.e. the curd is formed.

Cutting the curd is done using the lyre, but at home we can use a knife with a long sharp blade, cut the curd into squares and move the pot in a slight back and forth motion for 10 minutes, this helps the curd to drain.

Figure 30. Curd cutting

Figure 31. Curd heating

Heating the curd when a more compact curd and consequently a firmer cheese can be obtained by heating the curd together with the whey at a temperature of 50 °C and keeping it in a slow rotating movement for no more than 15 minutes.

Separation of the Curd now we separate the curd from the whey, using for this purpose the canvas. Let the curd drain for a few minutes, after which, taking the ends of the cloth, we make a kind of bag, squeeze it and force the whey out even more.

Figure 32. Curd separation

Molding the next step consists of cutting the curd and placing it in the molds, taking care to cover the molds with a cloth if plastic containers are used; if stainless steel containers are used, this is not necessary.

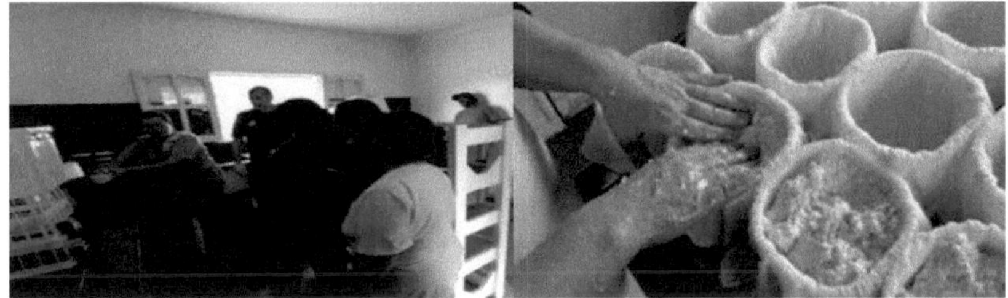

Figure 33. Curd molding

Pressed we put a weight on top and let it drain for a couple of hours. After that time,

we remove the cheese from the mold, separate the canvas and wash it to be used again, it is generally recommended to press with a weight equal or equivalent to four times the final weight that is expected to be obtained, it is also important to rotate the position of the cheese in the mold to keep the shape, well define the edges of the mold.

Figure 34. Cheese salting

The cheese is placed in brine (salt + water) which is prepared by dissolving one kg of common salt in one liter of water, this solution should be at a temperature not exceeding 15 °C. It should be kept in the brine for at least 3 hours, taking care that the cheeses are always submerged, in case they float the surface should be covered with solid common salt, it is extremely important to keep the cheese cold at all times.

Figure 35. Curd pressing

Once the recommended time for salting has elapsed, the cheeses go on to the last

stage of processing, where all the organoleptic characteristics (flavor, aroma, color, texture) are defined. For this to happen, they should be placed in a cold environment with temperatures below 15 °C, with low humidity and scarce air circulation. The total ripening time will depend on the type or variety of cheese to be obtained.

Figure 36. Cheese ripening

Once ripening is complete, the cheese is ready for

marketing and can be vacuum packed in wooden boxes, cardboard boxes, wrapped with paper, etc.

Figure 37. Packaged cheese and labeling

Conservation keep the cheeses in conditions until the moment of sale at low temperature and high Relative Humidity (to avoid weight loss).

Obtaining a healthy, safe and high quality product requires the control and monitoring of each stage of the production process, from feeding the herd, obtaining the milk, handling, processing and maturing of the cheeses.

Recommendations

Figure 38. Cheese preservation

- **For cleaning and disinfection** Utensils and all parts that come into contact with milk can be disinfected with hot water at 80 °C before use. A simple and water-saving method is to fill a tub with hot water and put directly into it all the utensils necessary for the treatment of milk. They are then removed directly from the hot water for use. During the cheese making process, the water cools down a little, but remains practically sterile. If visibly dirty and undisinfected utensils are put into the slightly cooled water, they should be replaced with fresh hot water.

- **For the addition of the ferment, allow the** necessary time for the lactic acid bacteria to be able to form lactic acid from lactose, the milk sugar. The appropriate rest is about 45 minutes.
- **After adding the rennet**, it is necessary to let the milk with the rennet stand for at least 45 minutes, during which time the container in which the coagulation is taking place should not be moved for any reason, nor should it be mixed.

- **Use of non-commercial rennet**

Extraction of rennet from kid

- Slaughtering a kid in a place sheltered from wind and earth.

- Remove and empty the rennet (or stomach), wash it with potable water and remove all the fat.

- Place the clean curd in coarse salt, covering it completely inside and out.

- Dry in the shade and protected from dust, for 3 days or more, until dry.

- Store in a cool place (refrigerator) and protected from light until use. It should not have unpleasant odor at the moment of use.

- **Use of rennet**

- Heat for 15 minutes 1 liter of fresh whey (from the last cheese making) to 70° C and then cool to 40° C (lukewarm).

- Place the serum in a wide-mouth glass bottle, well cleaned.

- Add the dried curd in small pieces and let stand for 2 or 3 days at room temperature, in a dark place. Then keep in the refrigerator. Use the preparation without adding whey or rennet again. It should not have an unpleasant odor at the time of use.

- **How to measure rennet titer**

- Place 1 teaspoon (1 cubic centimeter) of rennet in 100 cubic centimeters of milk (half a cup of tea).

- Leave to act for 15-20 minutes, during which time it should coagulate.

- If it does not coagulate in time, double the amount of rennet should be used to achieve the result, testing again.

- **For pressing**

It is advisable to use as pressing weight a value equal to that of the product being processed, for example, for each kilogram of cheese, use one kilogram of pressing.

- **Importance of salting**

It is suggested to carry out a correct salting because this is the way to achieve it:

- Improve flavor.
- Control the development of microorganisms.
- In some cases, if necessary, it helps to form the bark or shell.

Yogurt production method

Definition: Yogurt is the product obtained by acid fermentation through the action of Lactobacillus bulgaricus and streptococcus thermophilus, from pasteurized milk, with the addition of sugar, unflavored gelatin, permitted colorants and preservatives, whole milk powder, etc.

Raw materials and supplies

- Goat's milk: from healthy animals, without the presence of antibiotics as they will prevent the development of lactic culture microorganisms.
- Dairy cultures for yogurt (Lactobacillus bulgaricus and Streptococcus thermophilus) or a glass of plain commercial yogurt.
- Sugar.
- Unflavored powdered gelatin
- Fruit pulp
- Flavorings.
- Colorants.
- Preservatives such as potassium sorbate and sodium benzoate at a rate of half a gram per kilo of product.
- Whole milk powder.

Materials and Utensils

- Cultivation stove.
- Refrigerator.
- Heat source.
- Stainless steel pots, the largest of 50 liters and others of smaller size and volume.
- Thermometer (100 °C minimum).
- Metal ladles.
- Glass containers.

Yogurt production process

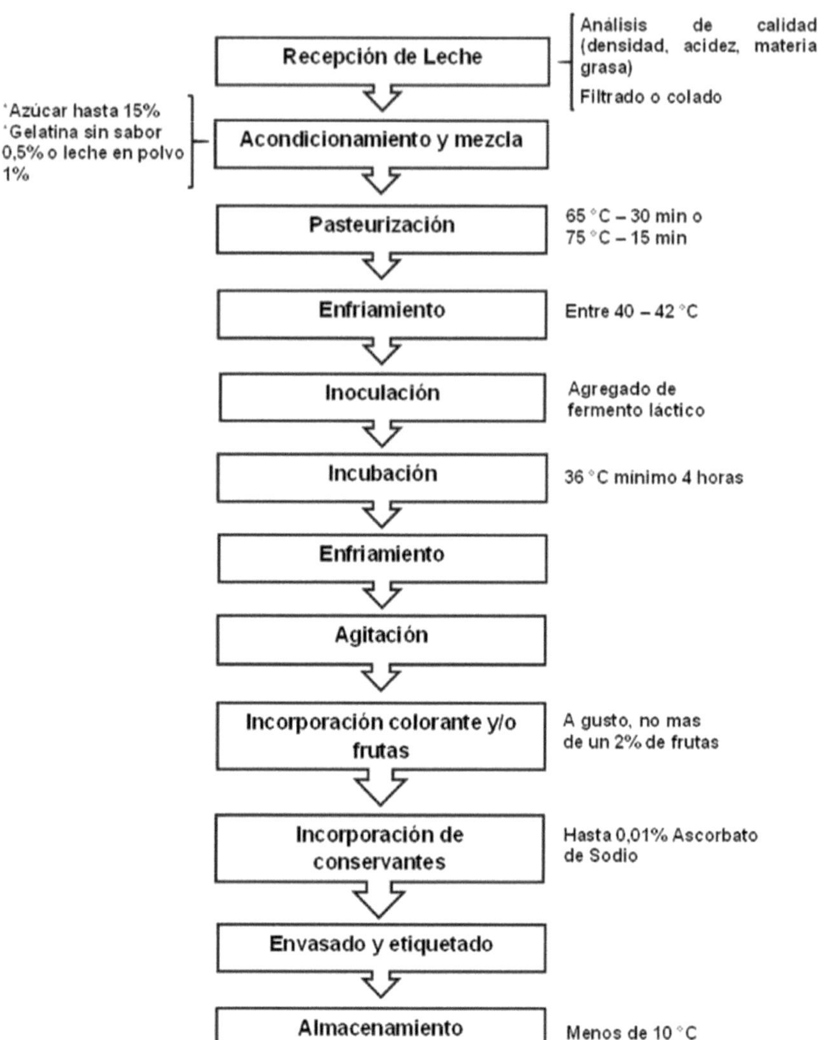

Figure 39. Block diagram of yogurt production.

Description of the stages:

Milk reception and quality control Milk is received and filtered by passing it through a clean cloth to retain any solids or foreign matter that may be present. The milk must be fresh so as not to alter the conditions due to an increase in the acidity of the milk by microbial action. Do not use milk from animals treated with antibiotics.

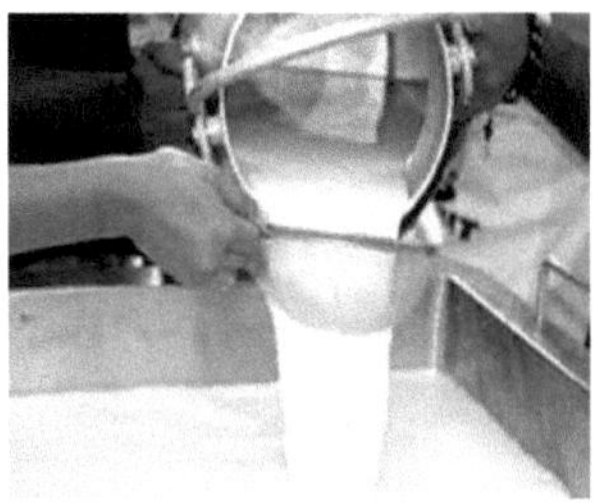

Figure **40.** Milk filtration

Figure **41.** Gelatin aggregate

Conditioning of milk For goat milk, add gelatin in quantities not exceeding 0.6% of the volume of milk, dissolving it first in cold water and then mixing it with the milk. Powdered milk can be added without exceeding 2% of the total volume of fluid milk, mixing well to avoid the formation of lumps; this can be done while heating the milk. If powdered milk is not available, add gelatin. It is also optional to add sugar at this stage.

Figure **42.** Pasteurization of milk

Pasteurization It is carried out at 70 to 75 °C for 30 minutes (slow pasteurization) or 90 °C for 10 minutes (fast pasteurization).

Cooling The milk should be cooled to temperatures of 42 - 45 °C (42 - 45 ºC).

Inoculation the milk is inoculated with strains of Lactobacillus bulgaricus and Streptococcus thermophilus as indicated on the package of the cultures, usually the temperature is 42 °C. Another way to inoculate is to use a glass of natural commercial yogurt, incorporating 3% of the total volume of milk, between 40-45°C, shake well to distribute the microorganisms uniformly

Figure 43. Inoculation of milk with microbial strains.

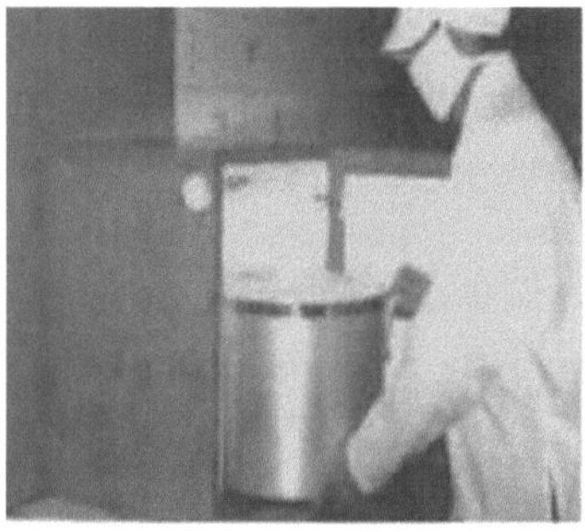

Figure 44. Incubation of yogurt

Incubation for 3.5 to 4 hours and at a temperature of 38 - 42 °C the milk must be kept, so that the microorganisms can develop and produce the desired characteristics in the yogurt. The end of incubation occurs when the product reaches an acidity of 75 to 80 degrees Dornic. If there are no instruments to measure acidity, the end of incubation will be determined by the person who makes the yogurt, with a visual examination, since the yogurt will form a compact gel, which when moving will have the characteristic movement of a jelly or empirically using a broom straw that when placed in the center should be kept in a vertical position

Cooling Once the time has elapsed and the acidity has been reached, the yogurt is cooled to a temperature of 5 °C by overflowing the water in the water bath or by placing the container in the refrigerator. The technological objective of cooling is to stop the action of the microorganisms, or a much more

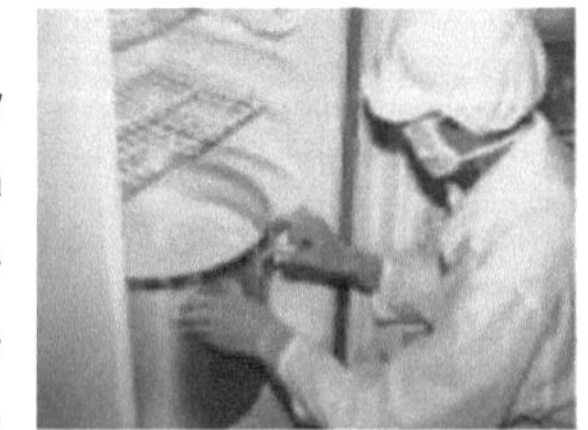

Figure 45. Yogurt cooling

acid yogurt will be obtained when the fermentation stops naturally.

Sugar incorporation and agitation Once the temperature has reached 5 °C or less, the sugar is incorporated at 5% of the content, which can vary up to 8%. Sugar can be added before pasteurization so that its granules dissolve well, but since the artisanal process does not have a common homogenizer, it is preferable to add the sugar at this stage, since, for whipped yogurt, the addition leads to strong agitation, which will eliminate the fat lumps (objective of homogenization), which is a defect of the final product

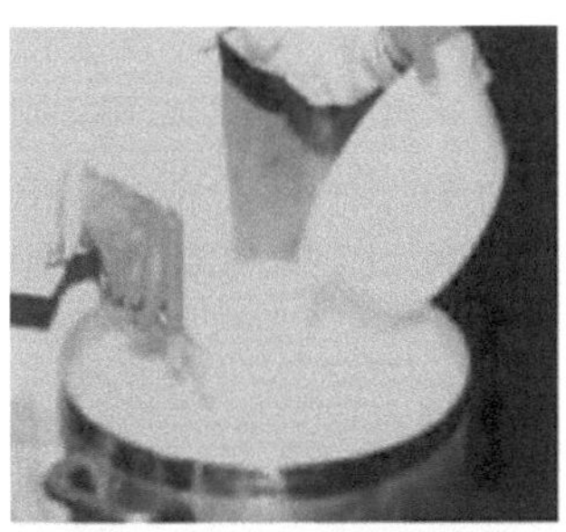

Figure 46. Sugar incorporation with agitation

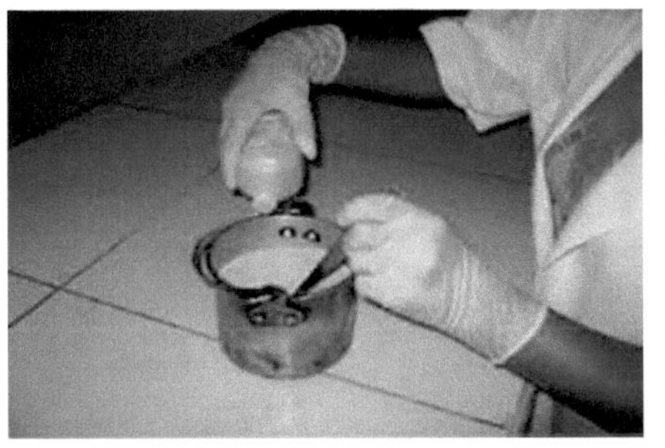

Figure 47. Dye incorporation

Incorporation of coloring and/or fruits Fruit pulp (10%) or coloring will be incorporated.

Figure 48. Incorporation of preservatives

Addition of preservatives Add 0.2 grams (beer cap) of potassium sorbate or sodium benzoate for every 10 liters of yogurt that was made. If the waiting time of the product for consumption is short, do not add preservatives.

Packaging and labeling Dispense yogurt (250 mL) into previously sterilized glass

containers (in boiling water at 100 °C) for 15 minutes
or at temperatures above 90 °C for 20-30 minutes,
cool by overflowing water, taking care that the cold
water does not come into direct contact with the
containers or they will break). Group by production
batches and indicate date of preparation, ingredients
and expiration date (approximately 5 to 7 days)

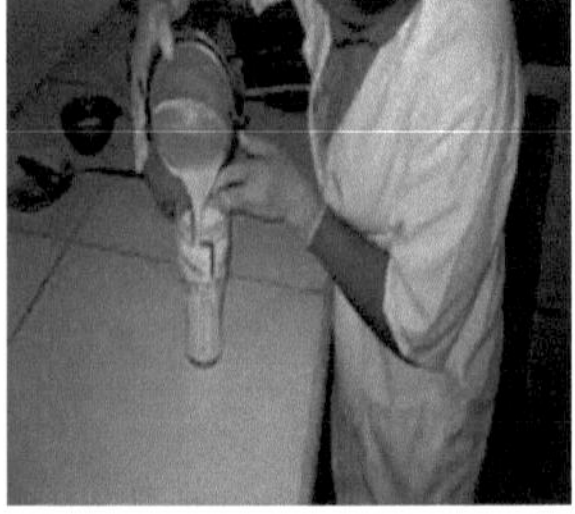

Figure 49. Yogurt packaging

Storage Refrigerate at temperatures of 5 °C, take
care not to break the cold chain.

Figure 50. Yogurt storage

Method of ricotta production

Preparation of Ricotta

Whey is a substance that remains after the curds are removed from the cheese. Whey is loaded with the proteins, minerals and vitamins that were in the milk. Cheese whey has a number of functional properties and a high nutritional value which allows it to be used in food use.

For the production of ricotta it is necessary to precipitate the protein solids contained in the whey and this is achieved by lowering the pH by adding a solution of an organic acid. Ricotta has a weak consistency, white color, odorless, sweet taste, although up to 2% of salt can be added, according to preference or taste.

Materials

-Stainless steel pot   or pan.
-Heat source
-Wooden ladle.
-Packaging
-Organic Acids (Acetic Acid, vinegar or lemon)
-Thermometer

Raw Materials

Curd whey obtained from the production of cheese, which must be kept at a temperature below 25 °C until it is used.

Process of ricotta production

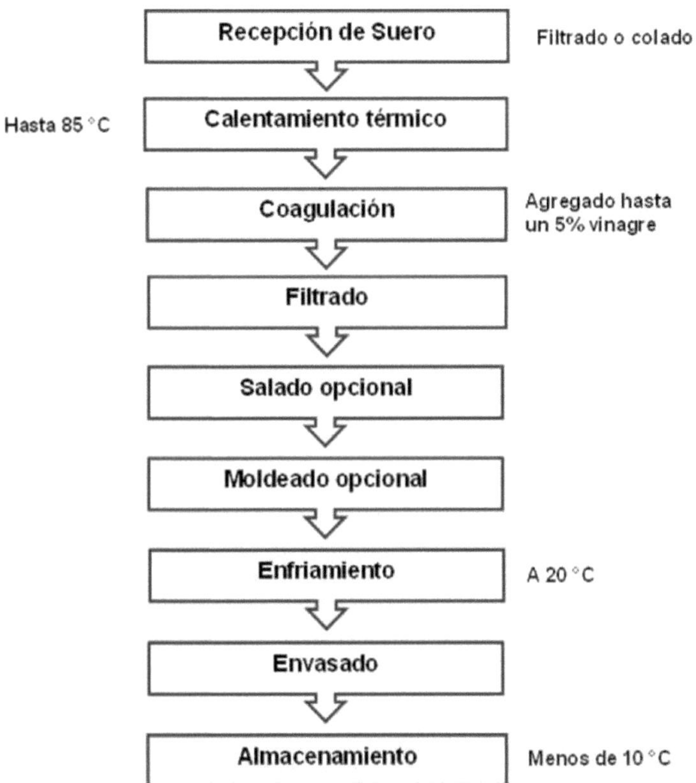

Figure 51. Block diagram of ricotta production

Processing method

Ricotta is a protein-rich food obtained from both whole milk and whey left over from cheese making. When it is made from whey, the steps to follow are described below:

Figure 52. Heat treatment of whey

Heat treatment Place the whey in a large vessel over medium heat using a thermometer. Wait for the temperature value to reach 85 °C.

Figure 53. Acetic acid aggregate

Coagulation When this temperature is reached, remove from heat, add 250 mL of 5% acetic acid solution (vinegar) for every 10 liters of whey, add the vinegar slowly.

Figure 54. Serum filtration

Filtering After precipitating the protein solids, small white flakes corresponding to coagulated protein will form and the coagulated protein is then collected using a filter to separate it from the rest of the remaining liquid or serum.

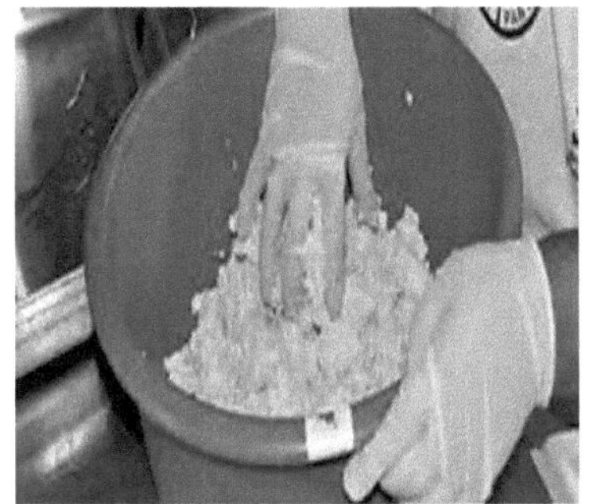

Figure 55. Curd salting

The curd is then allowed to cool and up to 2% salt is added, mixed by hand and left to stand for about 10 minutes for the salt to dissolve and distribute evenly.

Molding Molds of different materials and capacities (plastic, metallic, etc.) can be used. It is important to remember that it is essential to use a canvas cloth to prevent the ricotta from sticking to the inner surface of the mold.

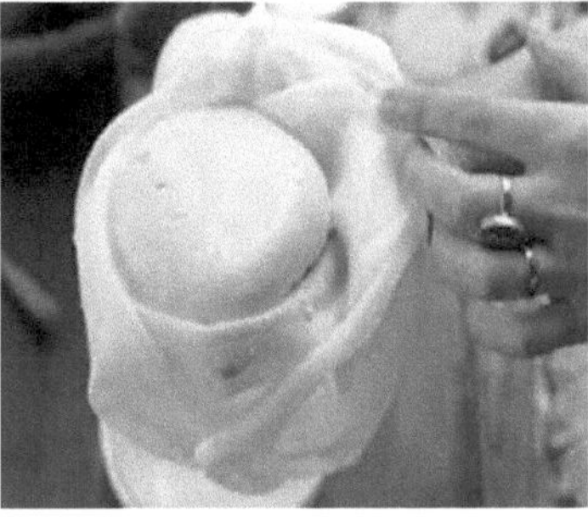

Figure 56. Curd molding

Cooling Rapidly cool to temperatures below 5 °C and above 1 °C, 3 °C is recommended, taking care not to freeze.

Storage The storage temperature is 3 °C and should be consumed quickly, as the product has a shelf life of no more than 5 days.

Recommendations

To increase the yield or quantity of ricotta, it is advisable to add fluid milk before coagulation. It is economically profitable to add up to 10% of fluid milk, for example, one liter of milk for every 10 liters of whey used.

I **want** morebooks!

Buy your books fast and straightforward online - at one of world's fastest growing online book stores! Environmentally sound due to Print-on-Demand technologies.

Buy your books online at
www.morebooks.shop

Kaufen Sie Ihre Bücher schnell und unkompliziert online – auf einer der am schnellsten wachsenden Buchhandelsplattformen weltweit! Dank Print-On-Demand umwelt- und ressourcenschonend produzi ert.

Bücher schneller online kaufen
www.morebooks.shop

KS OmniScriptum Publishing
Brivibas gatve 197
LV-1039 Riga, Latvia
Telefax: +371 686 204 55

info@omniscriptum.com
www.omniscriptum.com

MIX
Papier aus verantwortungsvollen Quellen
Paper from responsible sources
FSC® C105338
FSC
www.fsc.org

Printed by Books on Demand GmbH, Norderstedt / Germany